11.95

DATE DUE

MAR 2 6 1998

MAMMALS

ANIMALS IN DISGUISE

Lynn Stone

The Rourke Corporation, Inc.
Vero Beach, Florida 32964

PHOTO CREDITS
All photos © Lynn M. Stone

EDITORIAL SERVICES:
Penworthy Learning Systems

Library of Congress Cataloging-in-Publication Data

Stone, Lynn M.
   Mammals / Lynn M. Stone.
     p. cm. — (Animals in disguise)
   Includes index
   Summary: Describes how various mammals use ways to disguise themselves and fool other animals, including camouflage and other tricks with color.
   ISBN 0-86593-485-1
   1. Mammals—Juvenile literature. 2. Camouflage (Biology) —Juvenile literature.
[1. Mammals. 2. Camouflage (Biology)] I. Title II. Series. Stone, Lynn M.
Animals in disguise.
QL706.2.S76 1998
599.147'2—dc21
                                            98–6329
                                               CIP
                                             AC

Printed in the USA

TABLE OF CONTENTS

MAMMALS

Like other animals, **mammals** (MAM ulz) often depend upon their color to help them stay alive. The color of a mammal's fur is important. It can help the animal eat without being eaten!

A mammal's fur or hair is often a **disguise** (dis GYZ). A disguise helps a wild animal—or person—look like something it's not. Fooling other animals is one way that an animal can stay alive.

A fawn depends partly on its coloring for safety in the woodlands.

STAYING ALIVE

A mammal's disguise helps it hide. A lion's yellowish coat, for example, helps it blend into the yellow grass of its **habitat** (HAB uh TAT), or home.

By almost disappearing into the tall grass, a lion is well hidden. That gives the lion, a **predator** (PRED uh tur), a better chance at catching its **prey** (PRAY), or food. Without being able to catch prey, a lion could not live.

If a lion were easy to see all the time, its prey, such as antelope and zebras, would almost always escape.

The lion's straw-colored coat helps it sneak up on prey.

MAMMALS IN DISGUISE

Mammals like the lion can't truly disappear. They can make themselves hard to see.

Being able to look like part of their surroundings helps predators, like lions, to make kills. Some prey mammals, like mice, depend on their coloring to hide from predators. Sometimes the predators fool the prey. Sometimes the prey animals fool the predators. In this way, some members of each animal group stay alive.

Arctic foxes in winter white coats have a better chance of catching prey.

CAMOUFLAGE

Animal colors that match their surroundings are called **cryptic** (KRIP tik). With cryptic coloring, an animal can **camouflage** (KAM uh FLAHJ), or hide, itself in its surroundings.

Mammals almost never have camouflage as good as certain insects, fish, and **amphibians** (am FIB ee unz). That's mainly because most mammals travel easily from one habitat to another. Being camouflaged for just one place would not be helpful.

The sloth, though, is a mammal with amazing camouflage. Tiny green growths on the sloth's fur hide the animal in treetops.

Blending with its setting, a tiger cub practices hunting skills.

The short-tailed weasel, or ermine, wears a brown coat most of the year.

The ermine trades brown fur for white in the winter, giving it year-round camouflage.

NEUTRAL COLORS

Most mammals wear **neutral** (NOO trul) colors of brown or gray. In neutral colors, mammals don't call too much attention to themselves. A brownish-gray rabbit, for example, doesn't have perfect camouflage. But it can hide fairly easily. A mountain lion can travel in forest, desert, or rocky meadow and not stand out.

Deer, elk, and bighorn sheep are among the mammals with brown and gray coats. They are not perfectly camouflaged, but they also are not easy to see.

The earth colors of a ground squirrel help it hide among leaves, rocks, and soil.

TRICKS WITH COLOR

Some of the brightest coats are surprisingly good disguises. The tiger's stripes, for example, help it hide in tall grass or in woodlands. The spots of leopards and cheetahs help disguise them. Spots and stripes help break up an animal's form when seen from a distance.

Is solid black a good disguise? For black leopards and jaguars, it is. They hide easily in forest shadows.

A black leopard can hide itself in forest shadows.

USING COLOR

There is more to survival than wearing a disguise. The rabbit's brownish fur is some security, but only when the rabbit sits still. The disguise is lost when the rabbit runs.

Sitting still, ears flat, the rabbit looks like part of its surroundings. Running, the rabbit looks like a rabbit—and rabbit-eating animals may give chase.

Mammal predators, too, use their colors where they do the most good. A spotted or striped cat hides best when it visits a place of both shadows and light.

To stay safe, a cottontail rabbit can sit still and look like part of its surroundings.

SNOW WHITES

The Arctic lands of the far North are covered with snow much of the year. To match their snowy surroundings, the Arctic wolf and polar bear wear thick white coats all year. The Peary's caribou is whitish gray, much lighter than caribou farther south.

The Dall sheep of northern mountains is another white-coated mammal. The snowshoe hare, Arctic fox, and **ermine** (ER min) are white, too, but only in winter.

The snowshoe hare wears a white coat in winter and a brown coat in summer.

CHANGING COATS

Unlike most mammals, the snowshoe hare, Arctic fox, and ermine change their fur coats with the seasons. In spring, their coats begin to turn from white to brown, just like the land around them. By summer, these mammals blend in with their surroundings.

With shorter days in late summer, white fur begins to replace the brown. By winter, the white fur is back.

The fox and ermine use their camouflage to hunt. The hare uses its camouflage to hide from hunters.

Glossary

amphibian (am FIB ee un) — any one of a group of soft, moist-skinned animals that are usually born in water and usually become air-breathing land animals as adults; frogs, toads, and salamanders

camouflage (KAM uh FLAHJ) — the ability of an animal to use color, actions, and shape to blend into its surroundings

cryptic (KRIP tik) — that which helps hide, such as the colors of an animal that help it hide in its surroundings

disguise (dis GYZ) — a way of changing an animal's appearance

ermine (ER min) — another name for the short-tailed weasel

habitat (HAB uh TAT) — the special kind of place where an animal lives, such as a northern forest

mammal (MAM ul) — an air-breathing, warm-blooded, milk-producing animal

neutral (NOO trul) — having little color, such as gray

predator (PRED uh tur) — an animal that hunts other animals for food

prey (PRAY) — an animal that is hunted by other animals

INDEX

FURTHER READING:

Find out more about Animals in Disguise with these helpful books and information sites:

• Burnie, David. *Mammals*. Dorling Kindersley, 1993
• Carter, Kyle. *What Makes a Mammal?* Rourke, 1997
• National Audubon Society online at www.audubon.org
• Parker, Steve. *Natural World*. Knopf, 1994